Fishing in the Entomological Stream

Essays on Insects and Life by Students of Both

Edited by Jody Green, Thelma Heidel & Larry Murdock

Illustrations by Thelma Heidel

Purdue University 2007

ISBN 978-0-6151-5114-4

Published 2007 by Murdock/Lulu Press.

PREFACE

One of life's joys is teaching young people. In the spring of
2007 I taught a class at Purdue University with the nebulous
title "Fishing in the Entomological Stream". This was a one-
hour course of lectures and activities for graduate and
advanced undergraduate students. I offered it because no
other course in the University taught some key things I felt
students needed to know. I'm not talking technical content;
I'm talking about how one does science, how it really works,
ways of thinking about it, and sources of research ideas and
information. I'm also talking about alerting students to the
kinds of demands that will be placed on them, and about the
characters they can expect to meet (contend with, suffer from,
or work joyfully with) in their careers. I hoped that I was
offering them a friendly hand to help them as they climbed
their scientific ladder. I tried also to give them a few tips about
how to do science well, and tell them things I had learned the
hard way over many years.

My "Fishing" class was, I am proud to say, an unconventional
one. The reason I am proud is that these are the days of
grinding conventionality – what with benchmarks, strategic
plans, metrics, milestones and increasing administrative
accountability – all of which have no place in science, not in
true science, anyway, for true science is the province of

curiosity and wonder, not bean-counting. I tell my students –
only half jokingly – that Charles Darwin would never have
been able to earn tenure in a modern research university
because today's university would not have tolerated such an
original and self-motivated spirit as he. Imagine him writing
down his goals and benchmarks for the coming year! I am – of
course it must be obvious – old-fashioned, and I believe that
the best science is done in a state of fascination and
enchantment. When these are the driving forces, everything
else falls into place. They are the forces that have driven me,
anyway. I wanted to share.

One point I hammered on again and again in my lectures: the
importance of writing well. No skill, I said, was more vital. I
told the students that they would never learn to write really
well until they learned to think clearly. I told them clear writing
and clear thinking are inseparable; you can't have the first
without the last. Obviously, the best way to learn to write well
is to actually write – and edit the writings of others. With this
in mind I conjured up a class project: We would write a book,
with each student choosing her/his own topic for a chapter.
The only condition was that the student's story have an
entomological component; it made no difference whether it
was central or tangential. Each student's draft was edited by
the other students and by me, then revised on the basis of the

feedback. This edit-and-rewrite process was repeated several times as the semester went along. The students learned about writing and editing as a result – they learned that it is dog-gone hard (but possible) to write interestingly and well.

The book in your hand has three uses. First, it provides a sociological time-capsule. One-hundred years from now, interested parties can read profiles of the lives, values and thoughts of young students of the Purdue University Department of Entomology in the early Twenty-first Century. Maybe by that time no one will care, but I suspect a few will – even if it only be the authors' great-grandchildren. Second, it offers some interesting stories that are fun to read but which also shine lights into the hearts and minds of young people who somewhere in their youth acquired a passion for entomology – though there is one exception to this early passion, as you will see. Third, it gave the students themselves a chance to practice their writing and editing.

I am gratified to have been able to work with these students. Their fresh ideas and youth and hope warmed my heart. I don't know if they looked forward to coming to class, but I sure did. If Norman Rockwell were still around and painting, he would have found plenty of inspiration in the faces looking up at me during those fifteen class hours. As I said, one of life's joys is teaching young people.

Larry Murdock
Professor of Entomology
Purdue University
W. Lafayette, IN, USA

June 13, 2007

TABLE OF CONTENTS

ADVENTURES IN MY KINGDOM

By Alana Jacobson

eing the youngest of two children and the only daughter, I was coddled and protected more than my brother. I also got away with more than he did, which fueled the same jealousy that made him demand to my parents that they "take me back" the day I was born. My brother was five years older than me, and always at least four times my size. This, coupled with his jealousy, was dangerous for me. Lucky for me being the baby girl brought with it an entitlement: I was a princess. As a small child this title helped protect me from my brother and the multitude of dangers I encountered at home on my three-acre kingdom filled with animals and arthropods. Most of the time I avoided these perils, but on occasion I ran smack-dab into them and found myself living the life of a great action heroine braving treacherous cacti forests, fighting Cecil the samurai, or ironically falling victim to a baby arthropod who, like myself, just didn't want to get hurt. These dramatic situations usually proved quite comical after-the-fact, however, one of them could have been fatal.

Growing up wasn't always a bunch of roses. Actually, living in the desert meant there weren't any roses at all, but there were a whole lot of cacti. Even though these cacti represented one of the more obvious perils of my kingdom, I ran into them (literally) on numerous occasions. I remember riding my tricycle around one of the few concrete paths that surrounded one of my mom's cactus gardens. This had been one of my daily activities for several years, but one day a rock caught the back tire of my tricycle as I was rounding a corner, and I fell right into the arms of a Peruvian tree cactus. Lucky for me the Peruvian cacti had large prominent spines instead of the short, thin, blond ones that instantaneously meld into your flesh and inflict pain from invisible locations on and under your skin. The large size of the spines made it easier for Mom to remove them from my body. As she did, she warned me not to ride my trike around the cactus garden. Of course, I instantly returned to doing just that the minute she was done. I won't argue, it wasn't the brightest thing to do, but running into cacti was something that couldn't always be avoided. I accepted that this was going to happen on occasion. Besides, it

was the surprise attacks, not the known dangers that were more painful and insulting to me.

As I mentioned, many animals, including a flock of Chinese snow geese, shared my kingdom with me. A negotiation had been made between the geese and the monarchy; they were granted permission to leave the chicken-coop and run free in the kingdom during the day in exchange for providing pest control services. Even though their leader, Cecil, was known to have a bad temper, this arrangement worked well until the day he attacked me.

It was a hot summer day. The yard was full of water from being flood irrigated. I was in my swimsuit playing in the water as the geese foraged for worms and insects that had been displaced from their homes by the water. Cecil was in one of his bad moods, and the moment I was within striking distance he ran at me. Before I knew it I was in the middle of a kung-fu throw-down. Looking back, if I had been enrolled in tae kwon do instead of gymnastics at the YMCA, it might have been a fair fight. We were the same height, and had the same skinny arms and bony elbows. The difference was, I didn't know how to use mine. He did. He was obviously a five degree black belt and expertly used his wing bones as swords, swinging them with precision and striking me repeatedly all over my body. If that wasn't bad enough, he bit me too. Too surprised by the attack to even try and defend myself I did what anyone facing such a formidable foe would do; I found my exit and ran screaming and crying to Mom.

The King and Queen did not appreciate the attack or the bruises left on me. They wanted me to be able to play outside in the water during the summer days, but they also wanted the pest control services the geese provided. The political pressure from both the geese and myself led the monarchy to a compromise. A netted tent used during camping trips was erected, and a pool was placed inside to provide me with a goose-proof playing environment while the geese were allowed to run free. This worked well during the summer months, but the need to properly punish Cecil for his crime, and the worry of another attack never left my parents' minds. Due to this, the Royal Couple decided to include Cecil in our Christmas dinner plans. This eliminated the threat of more attacks, and allowed me the opportunity to bite back. The rest of our geese were released at one of the city's parks so they could terrorize other people's children.

Unfortunately, not all outcomes of my adventures were so sweet. Just like the plot in a typical drama movie, my parents tried to protect me from the very danger that happened to be one of my passions. My way of dealing with this undesired protection, although innocent, almost killed me.

My passion was horses. Even though we had four of them, I was not allowed to go into their pens or ride them because I was "too little". Regardless, I did realize that I could never saddle or mount a horse by myself; I was too short and weak. This, coupled with a fear of disobeying my parents sustained my obedience. It did not, however, constrain my imagination.

In my kingdom, I was the princess of countless horses of every breed and color. Every day I would pick out the one I wanted to ride and we would go racing around the yard, jumping fences, dodging trees (and cacti), chasing the wind, and riding off into the sunset. You might guess that all this came to life while I was astride my trusty stick horse, but he proved to be a disappointing substitute for a real horse, as well as an imaginary one. While riding my stick horse I couldn't run very fast without tripping over the stick and falling on my face. The only good thing about my stick horse was that I could keep him in my room and he wouldn't poop on the floor. Due to this, my stick horse remained in a corner of my room. Instead, a braided leather riding crop that hung on a hook in the tack room became the key to my horse adventures.

One evening before dinner, after deciding to saddle up my favorite black and white paint stud with the white bald face and blue eyes, I headed out to the tack room to get my riding crop. When I reached up to grab the crop off of the wall I received a shot of pain in my thumb that was so intense I ran, once again, screaming and crying to Mom. When I told her what happened she gave me an ice cube wrapped in a paper towel to place on my thumb, and told me to lie down on the couch.

After that there are few things I remember clearly. I recall my Mom calling to my Dad that we needed to go to the hospital, and having to wait for my brother before we left. I remember being strapped to a papoose board because I wouldn't hold still. I remember hating the papoose board and the people that strapped me to it. I remember riding home later and feeling hungry because we hadn't eaten dinner, and I also distinctly remember feeling sorry for all of the Indian children that had ever been carried in a papoose board.

Although I don't remember much of that night, the details of the complete story have been related to me many times by my parents. Before we went to the hospital, my brother, despite his contempt for me, bravely went out to the tack room, and captured what had stung me. It was a baby bark scorpion, known to entomologists and arachnologists as *Centruroides exelicada* (Buthidae). These scorpions are light-gold to golden-brown in color, and have one of the most potent venoms of all scorpions, and unfortunately for me, the baby scorpions' venom is even more potent than the adults'. These scorpions were abundant, and were frequently found inside the house on the tile floors or in the cool, moist kitchen sink. Everyone but me had already been stung and warned me of the intense fire-like pain that they experienced. When I was stung, however, I only remember a brief period of pain because after that I slipped quickly into anaphylactic shock. I was then carted off to the hospital where the nurses strapped me to the papoose board to keep me still while they poked me all over with a needle trying to find one of my tiny veins to give me an injection of anti-venom. I don't remember the poking, but according to Mom they were having so much trouble I was even poked in the bottoms of my feet! When they finally managed to find a vein and inject the anti-venom the convulsions and other symptoms subsided almost instantly.

To this day I still hate that papoose board, and find it strange that it was the white man and not the Indian children who called the Indians savages. I also hate shots, immunizations and needles, and want to run screaming and crying to Mom whenever I get them. I don't, however, hate scorpions. My first experience with scorpions left a lasting impression, which developed into a great respect for arthropods. My desire to study arthropods, however, did not manifest until I discovered entomology my junior year of high school. I never expected my little kingdom to lead me to where I am now, but I am thankful for it, and I look forward to the many adventures that await me.

ABOUT THE AUTHOR

Alana Jacobson was born and raised in a small kingdom in rural Phoenix, Arizona. She grew up frolicking with geese, chickens, dogs, cats, horses, goats, rabbits, and any other animal she could talk her parents into having. Her interest in entomology developed after being suckered into participating in the state entomology contest for FFA by one of her teachers at Carl Hayden High School's Agricultural Magnet Program. After graduating high school she went to New Mexico State University double majoring in Agricultural Biology and Spanish. Currently you can find her in one of the many sweet corn fields of Indiana, where she is conducting research for her master's degree, looking at insecticide resistance in the corn earworm. Horseback riding is her favorite hobby, but she is poor and cannot afford a horse, so she spends her free time frequenting the West Point Steakhouse (especially on Wednesdays - $0.50 draft night), and pillaging Indiana for fossils and geodes. She hopes to one day rule the world, and make more than $15,000/year.

"Reality is merely an illusion, albeit a very persistent one."

"The only thing that interferes with my learning is my education."

-Albert Einstein

LIKE FISH OUT OF WATER

By Jody (Aleong) Green

he first of many pictures of my sister, my brother and me on vacation was taken when my brother Jon was still in the womb. Our tall, skinny Mom looked ready to pop out a basketball. She wore a sweet youthful smile as she stood on a sandy beach, while Kelly and I posed on each side of the baby lump. I was seven and missing my front teeth; Kelly was four and cute like a doll. This was the first of many family vacations.

For years our annual Aleong Expedition coincided with Jon's birthday. Each one began exactly the same way: Our parents packed us up in our minivan and transported the whole clan from our home in the suburbs of Kitchener, Ontario to Orlando, Florida. It was the standard North American family vacation. As a result, for decades the only places we associated with the United States were gas stations in small towns and outlet malls along the highways that led to Disney World. Living in the Disney resort for a week was heaven for us kids, and who could blame us? We were, after all, in the land "where every Cinderella story comes true".

When teachers asked us to report on our summer activities, our trip to Florida was, by far, the most flaunted. Each fall we returned with shell souvenirs, a plethora of toys, colorful t-shirts, and pictures of ourselves with various Disney characters - all of which provoked jealousy in our classmates. It may be that Florida was freshest in our minds upon returning to school each fall, or maybe we just chose to forget "the Cottage" - our *other* annual excursion. Either way, none of us felt there was any need to brag about the cottage.

I'm not sure where Dad got his cockamamie idea of buying into a primitive lake cottage, but he somehow convinced my mother to financially partake in this venture. What was Dad thinking? Or more importantly, what was Mom thinking? She was (and still is) a woman who enjoyed the amenities that civilization had to offer, and had won the admiration of others for the way she could gracefully weave through crowds of frantic shoppers on Boxing Day. It would be a giant understatement to say she was out of her element at the cottage. Maybe Dad touted it as a great investment, I don't know. Maybe Mom just couldn't say no to her husband, whose childlike excitement at the prospect of a place on the lake convinced her to go along with buying a piece of wilderness. In any case, the cottage was a collaborative purchase involving Mom and Dad and Mom's brothers and sisters. They agreed that each family would inhabit the cottage for one or two weeks each summer. Our week was always in August, happily, the warmest time of year.

The cottage experience was a far cry from a Disney resort. It was there we kids learned what "roughing it" meant. Just as with our earlier Orlando trips, the five of us piled into our minivan, but instead of driving 24 hours to paradise, we drove five hours northeast to the middle of nowhere. On the way, we stopped at a small-town grocery store to equip ourselves with the necessities for the week including drinking water, biodegradable soap and shampoo, and, of course, toilet paper. We steeled ourselves for a week of monotone boredom in a technologically-impaired situation.

No road led to the cottage. It was accessible only by boat, unless it was winter and you dared drive across the frozen lake. Upon arriving at the marina where the cottage boats docked we found our little vessel and spent an hour struggling up and down a steep rocky incline transferring our supplies to the boat. It was a twenty-foot bow rider with seating for eight passengers. The ride to the cottage took thirty minutes, and after docking, we had no choice but to unload. The exertions required for the trip were exhausting; after all, we were young weaklings hauling twice our body weight to a place that was the exact opposite of Disney World. I can distinctly remember pulling up to the marina after sunset one summer and being frightened of the wind and the waves in the pitch dark, hoping and praying that our boat would miss the rocks and we would not all drown. With the vividness that only a young mind has, I imagined our small boat capsizing and decided which direction I'd swim in order to reach land quickest. I saw myself pulling my brother and sister to shore by their life jackets, too, and thereby becoming the hero of Mazinaw Lake.

The name Mazinaw means "painted rock" in Algonquin, the lake having been named for the many native pictographs on the face of Mazinaw Rock, the huge, towering cliff that rises tall out of the dark water. Mazinaw Rock could be compared to a wall in an art museum. The beauty of the artwork can only be experienced from a boat, and when you are slowly swaying and bobbing before it, everything is quiet and surreal. Among the petroglyphs is an engraved poetic tribute to Walt Whitman (who never visited the lake). It was chiseled by stonemasons in the early 1900's. That poem is the only recognizable writing on the rock. We visited this magnificent mural on the way over to the cottage, and along the way, admired the miles of blue skies that hovered over a thick and lush mature forest, the epitome of forest green. The cottage popped out of the canvas as we approached. It stood alone between two coves of small, private, sandy beaches. The cottage - which could more accurately be called a shack - was located on one of the fingers of Mazinaw, which by the way is the second deepest lake in Ontario. Its azure waters were surrounded by the greenest of forests and enormous, pinkest of granite rocks. From the cottage, we could hear our neighbors but not see them: The closest provincial park on the lake was named Bon Echo because voices and sounds echoed over the lake like a giant surround-sound system. In the morning, the eerie cries of a loon moving through the calm, misty surface resonated so much that it made the hairs on my arms stand up.

The cottage itself was nothing more than an old shelter made out of wood and rocks. The décor was a homely decoupage of different eras and unwanted items. The main room had three large windows that overlooked the lake and wrapped around the small kitchen. Two tiny bedrooms were in the back, but they may as well have been in the woods; they were dark, musty, and filled with the nighttime sounds of gnawing mice. Our parents slept in the less frightening of those two bedrooms, and the second bedroom remained empty. None of us kids had a desire to have a room all to ourselves – we knew there was safety in numbers. Accordingly, we slept on rows of air mattresses and sleeping bags in the main room by the heat of the wood stove. Power lines had somehow been strung through the wild woods to bring electricity. Electricity was Mom's saving grace. Because of it, she was able to spend her days at her sewing machine - which had been transported across the lake - and she could cook all our meals on the little electric range. Other kitchen appliances included an old refrigerator, an ancient microwave, and a beat-up toaster oven.

The most difficult thing for us soft-boiled suburbanites was the lack of plumbing. Water from the tap was siphoned directly from the lake, untreated, unpurified, and contaminated. We brought store-bought bottled water for drinking, cooking, and brushing our teeth, and we learned how to ration to make it last the whole week. Rather than shower, we were forced to lather up in the lake regardless of the temperature, which always seemed too cold.

Illustration by J. Green

Though the rest may sound bad, the scariest place of all was the dreaded outhouse. It was deep in the woods, a hundred paces behind the bedrooms. That awful shed was dark, stinky, and a nightmare to visit any time of day. Just walking back to it by the secluded dirt path of fallen pine needles, with millipedes and daddy longlegs crawling over our feet was enough to make any child pray for constipation.

There was nowhere to escape nature at Mazinaw Lake. Even inside the cottage, there was more nature than we needed or wanted. The cottage was alive at night, especially when we were supposed to be sleeping. Our presence didn't stop the mice and cockroaches from running around in the open. They weren't intimidated by the human giants striding around in their territory. One night, while changing in one of the bedrooms, I screamed when a mouse jumped up off the wood floor onto the curtains and started fabric dancing. My dad (my hero) rushed in and with one swift karate chop broke its back and sent it to mouse heaven. Another time, he attempted to shoot a mouse with a BB gun, only to miss repeatedly. He finally killed it by whacking it on the head with the gun barrel. I was a chronic insomniac and nights were long and terrible for me because I was the only one to witness the insect circus that performed while everyone else slept. Just going to the kitchen for a glass of water gave me chills. Cockroaches, beetles and earwigs crawled on the straw-textured walls and retro-laminate countertops. If I stayed awake reading, I sensed my personal space being slowly invaded by all the critters of the woods that were drawn to my light. One morning as I folded my linens, I spied a roach that used the space between my body and the blanket as its personal, warm, dark, crevice. I nearly had a heart attack at age thirteen!

At the cottage I was exposed to the scariest things in life – seclusion, deep and dark water, thick forests, and strange creatures. During the day, swallows living in their mud nests under the eaves dove at my head. At twilight, clouds of bats swooped through the skies above the lake vacuuming up the all-too-many insects. We spent entire days hiking in the woods, going nowhere except farther into isolation and always eager for evidence that white-tailed deer, moose, red foxes, and black bears lived in the area. At times the safest place was in the middle of the lake in a canoe, far from the land bugs, but even there seagulls coveted our picnic lunch and dopey horseflies buzzed around our heads, threatening to slash our tender skin with their terrible scissor-like mouthparts. In and around the lake were mussels, bullfrog tadpoles, a variety of fish, crayfish, giant dock-spiders, and the ever-present blood-sucking leeches. Prolonged exposure to these scary things forced me to face my fears and conquer them. For example, one time when we were fishing in the river, a leech attached itself to the bottom of my foot. Frantically waving my assaulted foot in my father's face, I shrieked, "Pull it off me!" My dad laughed and did absolutely nothing until I calmed down. We always knew we were lucky to have been taught by our father. A highly regarded veterinarian, he gave up his lucrative private practice for a federal government position so he could spend more time with his family. He was the most enthusiastic and practical science teacher we kids could have had in our youth, and we were forever changed because of his hands-on lessons. One of the lessons he taught us was how to hold a snake so that it was impossible for it to bite you. One day, after finding a family of harmless serpents resting within the root system of a massive, overturned stump, he proceeded to equip each of my hands with a stinking garter snake. I extended my arms out like a pro, at least long enough for everyone to acknowledge my bravery and take a picture. Then, as quickly as I could, I flung them into the lake. While I admit I was fascinated by the grace of their movements, which closely resembled living wavelengths as they sleekly skimmed the water's surface, I dashed for the woods in fright of snake revenge.

Although most of my childhood memories consist of panic and revulsion, the trip to the cottage is what I consider now to be the ideal family vacation. Being forced into week-long bouts of nothingness instilled creativity in our play, and having to continually interact with each other formed lifelong friendships, between the siblings of all generations. Our cottage was the perfect fit for my dad's personality, and he used it to shape *our* personalities. He taught us a love for nature, unstoppable curiosity, and a keen sense of adventure. Cottage living wasn't Mom's thing, so why did she suffer through all these crackpot adventures? The answer is simple: It was for us – It was because of her family.

*****

The cottage was eventually sold to a cousin and has since been renovated, upgraded, and modernized into something that I can't and don't want to imagine. I prefer to keep my memories of that odd sanctuary exactly the way it was when I was young. If it were possible to return to the cottage days, I would do it in a heartbeat, now that I'm finally able to appreciate the freedom seldom found in our lives today. Going to the outhouse no longer seems to be the worst thing in the world; it's more like a lost luxury.

ABOUT THE AUTHOR

Jody Green calls Kitchener, Ontario, Canada her hometown. She was originally born Jody Aleong, but lost her unique surname (to a color) after a marriage to an American food scientist. She has wild dreams of graduating from Purdue with her doctorate in May 2008, after 28 years of schooling in various disciplines and locations. Jody's main interest is Urban and Industrial Pest Management, and she performs her termite research experiments in the dead of night, when most normal people are sleeping. When not chasing undesired bugs, she can be found running outside, getting beefy at the gym, planning social events, tending to her spinach garden, playing fetch with her dog, Milo, or just walking around really fast doing things for other people. Her pet peeves are seatbelts that do not retract, human feet, and Indiana winters. Jody is a profuse and professional sweating machine who complains of the heat any time the temperature is above 25°C (77 °F).

"A little nonsense now and then is relished by the wisest men."

- Willy Wonka and the Chocolate Factory

FIRST KISS

By Christa Hardy

sat motionless, my eyes fixed on the rocking chrysalis before me. It twitched once, twice, and then a third time more vigorously. The tiny hanging cylinder rested for a second, allowing me time enough to concede a blink. I held my breath as it began to tremble again, for fear of interrupting the miracle before me. My jelly-stained hands were poised above the table top, the dry stickiness forgotten, along with the half-eaten sandwich I'd left on the deck. Awareness of my body, the kitchen I stood in, and even the table I leaned on vanished. As the sole witness to an indescribable wonder, every ounce of my attention was focused on each tiny movement of the chrysalis.

At the start of summer there were seven stalks of milkweed at the end of our driveway. Not being typical homeowners, my parents encouraged these tall plants to grow, mowing around this spot each Saturday morning. It wasn't until I discovered several plump yellow, white, and black-striped caterpillars on those milkweed plants that I began to appreciate my parents' unspoken intentions. Carried by gentle hands, we welcomed the still-munching bugs like honored guests into our home. A cardboard butterfly box with windows allowed us to observe the transformation from caterpillars to monarch butterflies.

The tiny, wiry caterpillars ate steadily, becoming larger, softer and plumper. One day, the zebra-striped caterpillars stopped eating and found a path to the lid of the paper box where they became trapeze artists, stitching a tiny silk-threaded hold to the lid. There they hung, with me watching spellbound, until darkness forced me unwillingly to bed.

I awoke the next morning to find one of the clinging caterpillars gone and a fluorescent green chrysalis in its place. I gazed in amazement at the tiny wrap of life held to the top of the box by only a thin black thread. Days passed without a change. I began to have doubts that there was still life in that green chrysalis. Not knowing the amount of patience required, I questioned whether the caterpillar had died. I was afraid to disturb the fragile environment for fear of tampering with things I ought not to.

For the second time a miracle happened while I slept. The bright green promise of life vanished overnight and in its place there was now a new profile: black veins and white spots beneath a transparent chrysalis skin. Hope was restored to my eight-year-old heart as the chrysalis became more lucent with each passing day. Not knowing what to expect, I had watched the tiny package of a butterfly every available moment for exactly two weeks. It was on a sun-filled Saturday afternoon, in mid-bite of a thick peanut butter and jelly sandwich, that I noticed a flicker of motion in the butterfly box.

The compressed and crinkled monarch had finally split its thin chrysalis walls. Slowly, uncertainly, each leg found a hold on the outer casing and pulled the rest of the body out. The crumpled wings unfolded as blood pulsed through the intricate vein system. I watched two pairs of newly unfurled wings beat slowly, cautiously, up and down. The slight breeze dried the perfectly patterned wings while I gazed in admiration at the vibrant orange shapes framed in black. The sharp contrast between white freckles and the rich black border of the wings held my stare for what seemed like hours. The monarch's delicate thread-like antennae arched perfectly toward tiny rounded tips and its soft black body held steadfast through each slow-motion spread of the wings. The miracle I had been waiting for was now resting gently on the ruptured chrysalis wrapping paper that it burst through to come into the world.

I took baby steps as I headed back out onto the deck, clutching my butterfly box. I ever so carefully lifted the lid and pushed my small fingers into the butterfly's world. The monarch stepped cautiously onto my finger, complimenting me with its trust. I lifted it into the fresh summer air, expecting that it would take flight and leave me for its journey.

To my great surprise my new friend did not relinquish its grasp, but instead climbed up my grey cotton t-shirt and rested on my shoulder. It then found my chin, and my cheeks. The monarch's movements were methodical, careful; each leg was thoughtfully placed as it made its way across my face, finally relaxing on the bridge of my nose. With one magical, orange kiss, it stretched its four paper-thin wings over my thick purple-framed glasses. My entire body was lit with a bright glow of joy and awe. In the small moment it took that butterfly to spread its wings, I became forever humbled before such an indescribable gift.

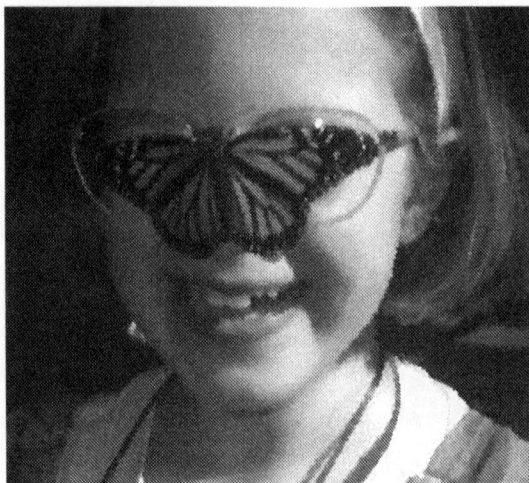

ABOUT THE AUTHOR

Christa Hardy is a junior at Purdue studying genetic biology and entomology. She enjoys warm weather and spending her free time outdoors. Originally from Ohio, Christa hopes to one day visit the Galapagos Islands. She sings loudest to country music and believes that a Chocolate X-Treme Blizzard is a solution to many problems. Around the entomology department, Christa is known for not having a mailbox. Christa believes…

"Those are a success who have lived well, laughed often, and loved much; who have gained the respect of intelligent people and the love of children, who have filled their niche and accomplished their task, who leave the world better than they found it, whether by a perfect poem or a rescued soul; who never lacked appreciation of the earth's beauty or failed to express it; who looked for the best in others and gave the best they had."

-Ralph Waldo Emerson

A DAY IN THE LIFE OF A SEVEN YEAR OLD

By Paul Marquardt

ne piece of yellow construction paper, a small bottle of Elmer's glue, and a plastic yogurt container, were my tools on that sultry summer day. I can't specifically remember if I had a clear goal in my head when I woke up, but by mid-morning I suspected it and when the clock struck 12 noon, I was dead certain of the path my day would take. I was seven years old at the time, living in Emporia, a medium-sized suburban town in east central Kansas. Thanks to being the son of a high school biology teacher and a Lutheran minister, I enjoyed certain perks during the summer months. Since my father came home around noon for lunch and my mother was off for the summer, my brother and sisters and I were rarely allowed to sit idly inside wasting a beautiful day playing video games. Not only did this not bother me, it gave me the freedom that any kid my age needed. It freed me to explore my neighborhood and get into as much mischief as I chose. This usually involved muddying up my new shoes or stealing melons and peppers from our neighbor's garden, but on this particular day I had made my way over to a favorite hang-out spot: "the creek".

Truth is, the creek was more like a cement algae slick than a natural waterway. It rarely had enough water in it to warrant the name. It was actually a drainage ditch built into the city sewer system to help drain runoff during summer thunderstorms. At one end it disappeared into a pair of 4-foot diameter cement tunnels covered by a steel grate. This end of the creek was a great pool of muddy, debris-laden water after thunderstorms because the steel grate soon became plugged with leaves and sticks and the water backed up behind it. The pool was off-limits to us kids by parental decree, yet that didn't stop us from wading into the dirty water after storms to collect our small aluminum-can sailboats as they slowed to a halt after a race down the flooded drainage ditch. I wasn't heading to this spot on that day though; I was headed to the opposite end of the creek.

The opposite end is where the concrete ditch gave way to a small, natural waterway that flowed under a barbwire fence. That fence separated the neighborhood from Interstate 35, which split Emporia in half. The great part about this small waterway was the large number of critters that flourished in the usually stagnant water. On this day I was engaged in one of my favorite past times: catching crawdads. I waded into the pool and began flipping over slime-covered stones to reveal terrified little crustaceans darting to safety in every direction. The larger

crawdads moved slower and were easier to catch. These were
the ones I was really after. I usually let them go after I caught
them, but today I intended to take some back to my garage to
keep in a small fish tank. Soon satisfied with my haul of five or
six 2-inch-long crawdads in my yogurt cup, I made my way
home. There I dumped my catch into my fish tank full of hose
water. Given my attention span (that of a seven year old), I
then took my yogurt cup outside to our pussy willow tree in
the back yard where I scampered around catching butterflies,
bees, wasps, and any other insects that had the misfortune of
crossing my path.

My goal for the day finally began to take shape. It dawned on
me that I should make an insect collection with what I had
caught in the backyard. It wouldn't be anything fancy with
pins and flattened butterflies or stuff like that, only a simple
yellow sheet of construction paper and Elmer's glue. As I
began gluing the still only half-dead insects to the paper I had
the bright idea that I could also include my day's catch of
crawdads. After gluing down a large bumblebee and a small
white butterfly, I fished out my largest crawdad (also only half-
dead) and dried him off before squirting a large glob of glue
onto the paper. I pushed the unfortunate crustacean down
into the glue, satisfied that my work was done, placed my
construction paper collection on top of my father's steel table

saw before heading in for dinner. As you might expect, given that I was only seven, I promptly forgot about my little project in the garage and after dinner headed to bed already thinking about tomorrow's mischief.

I was never allowed to sleep in much past 9 AM in the summertime, and the next morning was no exception. In fact, Dad woke me unusually early, before breakfast. He said he had something important he wanted me to see. Still in my pajamas, I put on my shoes and walked out into the garage with him. What I saw has been a part of Marquardt family history ever since.

My eyes instantly focused on my yellow sheet of construction paper on the otherwise spotless table saw. My curiosity about the unusual early wake-up call turned to fear and horror. The construction paper was no longer a pristine collection of Kansas arthropods at all. Instead, it was more like a B horror film. The Elmer's glue had dried unevenly, allowing most of the poor insects to rip off one or two of their legs before ultimately dying because they were unable to completely escape the sticky mess. This, though, was not why my obviously irritated father had dragged me out of bed. The problem was the big crawdad. He was able to accomplish what the smaller insects could not. He got completely loose. The spot where he was supposed to be affixed was now merely the beginning of a long trail of glue. It wandered off the paper, onto the steel top of the table saw and down onto the cement floor. There at the base of the saw were the sickly remains of this prize piece of my insect collection, its tail and legs covered in Elmer's glue.

Before my wedding, almost 2 years ago, this was one of the first stories that my parents told my in-laws at the rehearsal dinner. It started this way:

In-law: "So what's Paul do?"

Parent: "He's in graduate school at Purdue studying entomology."

In-law: "Entomology? Isn't that bugs?"

Parent: "Yeah, he's always been interested in them from an early age. Let me tell you about the time he was seven. . ."

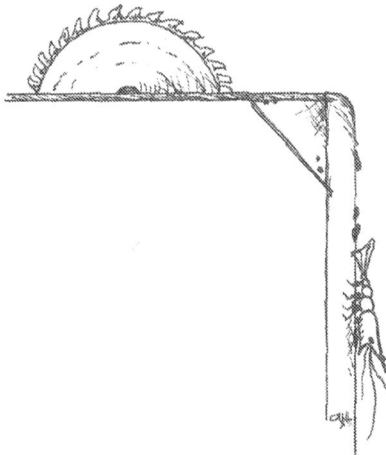

ABOUT THE AUTHOR

Paul was born in the small town of International Falls, Minnesota. He was lucky enough to be born just south of the Canadian border, half a mile inside U.S. territory. The vast majority of his life was spent in Emporia, Kansas, before attending Pacific Lutheran University in Tacoma, Washington, where he received his B.S. in biology in 2003. After that he wandered in and out of a few different jobs before moving to Seattle to pursue a job as a pseudo-vampire (night shift at the Puget Sound Blood Center). After two years at the blood center Paul moved to Indiana to begin his second tour of Midwest life as a graduate student at Purdue University. Paul loves to be outside and will try any outdoor activity at least once. He's an accomplished scuba diver, and is proud to have visited the 1st and 2nd largest barrier reefs in the world. Paul loves to have fun, drink beer, and watch pretty much any sporting event available whether live or on TV. He's an avid fan of the Chicago Cubs, the Kansas City Chiefs, and the Kansas Jayhawks, and is delusional, according to his loving wife Amber, to think that any of them will ever win a championship.

"All time is all time. It does not change. It does not lend itself to warnings of explanations. It simply is. Take it moment by moment, and you will find that we are all, as I've said before, bugs in amber."

-Kurt Vonnegut

DAD'S ADVICE

By Janice van Zee

hen I was asked to write about my "entomological" experiences, I had to ask myself if I had ever had one!?! To tell the truth, I would love to have had an enduring dream of becoming an entomologist, collecting insects, studying them, being mystified by their beauty… but I simply did not have such dreams… It is just not part of my personal history…Nevertheless, this is my story.

I was born in a small town in the Southern part of Brazil. Despite the fact that we lived in town, we kept our family farm, and many of my childhood memories were of weekends spent there and of all the silly things that a little girl with two older sisters could do. My grandparents built and lived in that farm house most of their lives; my father was a young boy when the place went up. My two older sisters were born there too. Thus the whole place has a lot of family history and memories for me. Unfortunately, after my grandparents passed away a few years ago, my family decided to demolish the house and construct a new one to modernize, I guess you would say. I still own the old house, with its water-well right in the middle of the yard and grape vines surrounding the house – in my memory. I am glad that I was able to enjoy that treasured old place, since the building… though not the memory… now is gone forever.

For most people, farms are places where you work hard. Fortunately for me, our farm was just a fun place to climb trees, swim in the lake, and play in the stream that crossed our land. I also liked to run around with the ducks, chickens and all sorts of farm animals. My two favorite things to do by far were fishing with my father and riding my horse. Nothing in the world could be better than that! I still do those things whenever I go there.

Because my parents did not live at the farm full-time, we only raised cattle and bees, neither of which required everyday attention. However, once in a while things got busy around there, especially on the days that my father needed to harvest the honey. We had Africanized honey bees, which are famous for their aggressive behavior. On those honey-gathering days we were "gently forced" to stay inside the house. You can imagine how frustrated I was for literally being locked inside a small farm house without television, computers or any other sort of technological entertainment. Needless to say that video games, X-boxes and I-pods were simply not present in my wildest dreams; for all practical purposes they did not exist. Even so, being incarcerated inside that little farm house was how my two older sisters and I spent some of our most precious weekends. For us kids, those were the sourest days; but for my dad they were the sweetest. Why? Because it was then that he harvested his golden honey.

We had bee-boxes all over the farm, most of them clustered in an area that had been kept natural. This is a common practice in the south part of Brazil to keep portions of our farms undisturbed, where we can still to this day find several different species of flowers, trees and animals. No crops, animals or tree logging is practiced there. We even try to avoid contact with that area to help preserve the wildlife. The natural areas are sanctuaries inside our farms, like having little pieces of heaven on Earth.

Illustration by J. van Zee

In my eyes, my brave father was like Super Man: he would go touch those boxes and wake up hundreds of thousands of mean bees, who could if they got angry, sting him to death. For some reason that I still can't understand, they rarely stung him, despite the very limited protective gear that he used.

He collected the honey, cleaned the boxes, prepared them for the next season and then came back home with lots and lots of sweet, sticky, runny combs of honey. We always had as much honey as we wanted around the house, but nothing compares with the delicate flavor of freshly harvested combs not yet prepared for consumption, especially the ones that your dad emphatically told you not to touch. Those were the very best! And of course we would sneak into the kitchen and "gently borrow" some of that precious manna. "Gently borrow" because you do not steal from your parents, you just anticipate the gift!

My mom used to tell me that if I ate too much honey, my belly was going to explode; and if I ate honey now I would not get any candy later. She also gave some more realistic advice: one day you will eat a honey comb with a bee inside, it will sting you and then you will see how good it tastes!!!

And that was exactly what happened to me. Mom was right! I was almost 4 years old and as silly as a child can be, when a mean bee hiding in a forbidden comb, stung me right in the roof of my mouth. Let me tell you, it really hurts. I had a big swollen bump inside my mouth and could hardly talk or eat for days. The doctor told my parents that the only thing he could do was give them even more strict advice about keeping me away from the honey. My sisters of course got a good laugh at my pain and suffering, and my mom, well; she got the reaffirmation that Mom always knows best…and for some curious reason she made sure to keep reminding me that she told me so all these years.

Illustration by J. van Zee

After that episode, I decided not only to leave the bees alone, but also to follow my dad's advice (at least sometimes!).

ABOUT THE AUTHOR

Janice van Zee was born in Pelotas City, Rio Grande do Sul. As a young girl, her family moved many times throughout Brazil, which included living in five different Brazilian States for various periods of time. Janice attended college in Pelotas with a major in Biology before accepting an invitation to continue her education and earn a Master's degree in the United States. Four years ago, while studying English in Kansas, she met an American soldier named Jeremy. He helped her perfect the English language, and they were married shortly after he returned from Iraq. Janice is currently working on her Ph.D. at Purdue University with a focus on medical entomology. Janice enjoys doing crafts, embroidery, painting, drawing, running, biking and dancing. She hates vacuuming and has therefore made that Jeremy's job. Janice will always have a cherished love in her heart for Brazil and encourages everyone to visit her beautiful country.

"Where there is love, distance doesn't matter."

-Mata Amritanandamayi

DADDY'S LITTLE GIRL

By Jessica Platt

hen I was a little girl, my dad worked at night. Thanks to this, I was able to play with him all day long. You could, with exact accuracy, say that I was "Daddy's little girl." Instead of playing with dolls, my days were filled with helping him. I assisted him in mopping the floors, with my miniature mop in hand, and even served as his personal assistant, working side by side with him, as he fixed miscellaneous things throughout the house. No matter what we did we always had fun! Between outings to the zoo, and my dad's place of choice, the library (probably because it was free!) every day was an adventure. But, among all of the different things that Dad and I did, I think that we would both agree that our favorite was to go fishing.

The day before one of our big expeditions, we dug worms for bait. This was something that I delighted in! All day long we scratched and spaded in the flower beds around our house, trying to find as many worms as we could. Digging for our bait brought to light many other fascinating critters. Some would fly away as I tried to capture then, some ran rapidly on spindly legs, and some just were squishy, but no matter what they looked like, I would always ask my dad what each one was. Growing up on a farm in Kentucky, Dad seemed to know a lot about insects and wildlife in general. He explained to me that the roly-polys rolled up into tight little balls just as soon as I picked them up only because they were scared of me. He also would always try to teach me the "bad bugs" from the "good bugs."

Among other things, he taught me a poem about lady bird beetles that went like this:

> *Ladybird, ladybird fly away home.*
> *Your house is on fire, your children all grown.*
> *Except little Nan,*
> *Who sits in a pan,*
> *Weaving her laces as fast as she can.*

He explained how the lady bird beetles were good bugs that ate other insects that could hurt our plants, and that this poem was made up to tell the lady bird beetles to leave the crops when they set them on fire.

One insect in particular I will always remember is the cut worm. Now I know that what Dad referred to as the "cut worm" was actually a Japanese beetle grub, his message was loud and clear: this was not an insect that he wanted to have around. Every time that I unearthed one of these "worms" I chopped it in half with my little shovel. I know this may sound a little violent, but we only did it for the sake of the flowers.

After collecting all of the fishing worms we could find, Dad and I put them in an old coffee tin filled with a mixture of dirt and coffee grounds so they would have something to eat until morning. We would then get out our fishing poles and tackle boxes. We were both so anxious for Mom to get home from work so we could show her all of the worms we had found.

The next morning, before Mom reluctantly had to leave for work, she would always make us bologna and liver cheese sandwiches on white bread, mine without crust of course, but with lots of mustard. To most people, this would probably sound like the most disgusting sandwich ever created, but, because I thought Dad was the coolest person in the world (which is still true to this day!) and this was his favorite sandwich, naturally, this was *my* favorite sandwich too…that is, until I actually learned what bologna and liver cheese were made of!

As we fished for bluegill, Dad pointed out all the different birds that we saw. He called my attention to the animal tracks that were around the bank from which we fished from and again, taught me about every little insect that I saw until I could point them out on my own. He would tell me horrible jokes, jokes that only a daughter would think were hilarious. Dad really taught me to appreciate all of the small things in life.

At the end of our fishing expedition, we would take all of our fish home. While my dad would clean our catch, I would help Mom get everything ready for dinner. We would laugh and talk about how much fun Dad and I had. Nothing was more enjoyable than getting to have a family dinner with our freshly caught fish!

ABOUT THE AUTHOR

Jessica Platt was born on December 29, 1984. She grew up with an amazing family in a southern town in Indiana called Jeffersonville. With her sisters being twelve and sixteen years older than she, she is definitely the baby of the family, and has loved every minute of it! She has always enjoyed doing anything outside with her family and friends whether it is attempting to play sports, camping or fishing. She also is always up for going shoe shopping (or any other type of shopping for that matter) with her mom. When she set off to Purdue in August of 2003, it was no surprise to her proud parents when she decided to change her concentration from Pre-Veterinary medicine to Entomology. If she is not in her lab, at Smith Hall, attempting to identify invasive bark beetles, she can usually be found going on a "Den run" to the Discount Den with some of her friends for a Den Pop! Upon graduation at Purdue in May, 2007, she will continue to study entomology at the University of Florida, working with a gall wasp where she will have a lot of opportunities to do extension and outreach activities (two of her favorite aspects of entomology)…after that, who knows!

> *"Sing like no one is listening,*
> *Love like you've never been hurt*
> *Dance like no one is watching,*
> *And live like its heaven on earth."* - *Mark Twain*

LIFE WITH FRANCES

By Thelma Heidel

had known Frances for about a week when I decided she needed to leave. I was tired of people talking about her. I was tired of her demands. She tried my patience and my resolve. Yes, maybe she had made me a better person, but that wasn't something I had asked for. It was simply time for her to go.

Frances was a hurricane.

I grew up a Midwestern girl with Midwestern weather. I was accustomed to tornado drills in springtime, shoveling piles of snow taller than me, and temperatures dropping beyond 20 degrees below zero in the wintertime. I remember staying home from school once for a week because of an ice storm; it was great fun. I was never afraid of extreme weather -- until Frances came along. I now found myself in a completely new situation. I was being embraced by something huge and frightening, something I really didn't think I'd ever be running from, and that something was Frances.

Two days had passed since Frances first introduced herself to central Florida. I knew this day would be like the day before, yet another day trapped indoors hiding from the hurricane winds and pounding rain. I stood at the large bay window looking up at the swirling grey clouds. We had been trapped in the house for a full 26 hours already, and Frances didn't show any sign of relaxing any time soon. "We" consisted of my friend Kim, her mom, three dogs, a cat, and myself. The house was a small single-level place outside of Gainesville, Florida, and in such small quarters, the aroma of "dogginess" that permeated every corner of that house was inescapable. We had been without power for nearly two days already, thanks to the first gusts from Frances. We ate food from cans and spent the day listening to the battery-operated radio. We watched the rain pound down and the winds tear through the treetops. We heard the crack of trees coming down outside and were helpless to do anything except hope nothing came crashing through our roof.

I quickly tired of Hurricane Frances. I couldn't go outside; it was too windy for that. Playing pool by candlelight had lost its charm early on. The radio, our lifeline to the outside world, comforted us with its ceaseless chatter of repeated weather reports. But even the radio wasn't enough to stop me from feeling a bit stir crazy. I desperately needed fresh air and a change of scenery, but Frances wasn't ready to let me go outside yet.

Then it happened. After nearly two full days of blasting winds and beating rains, Frances let up a bit. The sky turned from dark menacing grey to something lighter; I could finally see the low clouds racing by overhead. The chance to escape my prison had arrived. I hurried out to investigate and see firsthand the damage Frances had caused, and I drew a quick breath as I looked around. Tree branches were strewn everywhere around the house and backyard; a fallen tree had missed the house by mere inches. Not all of the neighboring houses were left so unscathed, however; broken windows, missing roof shingles, and downed trees were just a few of the mementos left by Frances. I knew it would take many days and a lot of work to bring this hurricane-stricken world back to some semblance of normalcy.

I stood in the backyard of the house that had served as both a safe haven and prison during the storm and wondered where all the living things had gone. Had they made it through Frances' winds and gusting rains? No birds were flying or singing. No insects buzzed about. Except for the tree rustling above, I could hear no sounds of living things. Yet I knew they were out there somewhere; I simply needed to be patient.

Slowly the animal life began to reveal itself to me. I noticed that the small koi pond in the backyard was teeming with frogs. They seemed unaffected by the storm. The dogs, as excited as I to finally be outside, also found life in the form of a wet chipmunk hiding underneath the shrubs beside the house. They chased that poor chipmunk around the yard again and again before he escaped beneath the neighbor's fence.

While observing this celebration of returning life, I began to feel a tickle on my bare feet, something like a stray grass blade rubbing against skin. At first I paid it no attention and continued to watch the happy frogs frolicking in the pondwater. But being a ticklish person by nature, I soon could no longer ignore the sensation. When at last I looked down, I was appalled at what I saw. My feet were crawling with tiny reddish ants! There must have been millions of them. They were between my toes, on my feet, under my feet, and all over the rocks I was perched upon. It looked like I was wearing a pair of red socks. Fear and loathing began to build inside me because I knew immediately what they were. Fire ants!

*****

Here I must back up a bit. When Frances arrived, I had been living in Florida for six months. My time there was the result of a combined desire to escape from the Midwest and the need for a job after college graduation. Moving to Florida only added to my appreciation for the warm tropical sun, palm trees, ocean beaches, and the fun insects and other creepy-crawlies of warmer climates. One insect in particular had continually fascinated me. It was the little ant that had become the bane of the southeastern United States, the people-biter that prevented folks from sitting on the lush green lawns of Florida, and the annoyance that I had – until this point in time – avoided contacting: the infamous fire ant.

*****

As I surveyed the invasion of my personal space, panic rose in me. I knew that the instant I moved, each tiny ant would attack my stark naked feet simultaneously, and I was very aware of the wounds fire ants caused. It just so happened that Kim's job was researching fire ant control, and she had the scars to prove her work. What unnerved me even more was that I had once witnessed the consequences of a fist being shoved into a fire-ant colony, and what I saw wasn't pretty. At the moment, however, I was most worried about the long-lingering blisters. I needed my feet absolutely flawless because in five days I would be embarking on one of the craziest and most physically challenging endeavors of my life, the Ironman Triathlon. I had been training for this race for the past six months. I wouldn't let a few fire-ant blisters stop me now, but I knew that even the tiniest of blisters could have devastating consequences during this 140.6-mile race. Thoughts of the dreaded DNF (Did Not

Finish) flashed before my eyes as I tried desperately to stand still.

Then it happened. I moved. I really tried not to. It was the slightest twitch but movement nonetheless, and the pain was instantaneous, like thousands of needles jabbing into me all at once. It was then I began the wild dance of the fire ant. Chaotic arm flailing, frenzied foot rubbing, and erratic hopping from foot to foot were just a few of the steps. (Maybe you have seen it, or even experienced it, but I really hope not!) Not wanting to dance anymore, I declared war on those ants. My war plan: to remove them by any means possible. I couldn't decide which foot to save first, so I sat down on my behind and picked viciously between my toes; it was there that the bites hurt most. Those tiny ants held on so hard. If they weren't clamped onto my feet at that exact moment, I think I could have admired their extreme tenacity.

Within a few minutes my war with the ants was over. They were gone, driven back by my maniacal and haphazard counterattack. The immediate pain went with them, but I knew the blisters would inevitably follow. I didn't blame the ants for their attack. They were simply checking out the aftermath of the storm, as was I. I just happened to stand on their home at exactly the wrong moment, and they countered my intrusion in the only way they knew how. All I could do now was wait: for the blisters, for the power to return to the house, for Frances to pass, and for life to return to normal.

Another day passed before Frances moved on for good, and her departure was greeted with a sigh of relief from everyone trapped in that small Gainesville home as well as the rest of Florida. Frances definitely left her mark before moving out of the state: houses destroyed, trees down, rivers flooded, and cars overturned. Physical objects, however, are nothing compared to life itself, and we had survived. A few days passed before I could escape the debris and destruction of Florida and fly to Wisconsin to compete in the Ironman Triathlon, and proudly, I can say I finished the Ironman, despite the fire ant blisters I carried on my feet that day.

ABOUT THE AUTHOR

Thelma Heidel is the oldest of three children born to dairy farmers in Random Lake, WI, where she grew up and raised every farm animal imaginable. Her original college plans were to become a veterinarian, but college opened her eyes up to new ideas. Thelma received her B.S. in plant pathology and entomology at the University of Wisconsin–Madison before choosing a career path in research. Before beginning her Masters she had a year-long stint in Florida working in the "real world" for the company Syngenta. She then realized she wanted to be a life-long student and began graduate school in the Biological Control Lab at Purdue University's Department of Entomology. She is known there as "Aphid Girl" because she can spot aphids from unusually far distances. Thelma loves traveling and adventure. She frequently visits the Philippines, where she has family and has studied abroad. Thelma is a certified SCUBA diver, avid rock climber, highly trained triathlete, and accomplished pianist. She hates cold weather and sometimes still wonders why she chose to go to grad school in a cold climate.

"All life is an experiment. The more experiments you make the better."

- Ralph Waldo Emerson

WHAT HAPPENS IN ZAMORANO STAYS IN ZAMORANO!

By: Diana Castillo

ot too long ago, when I was a freshman in College, I had little visitors come to my closet. You will understand why if you keep reading.

The school I attended is called Zamorano University. It is located in the very tropical area of Central America, more specifically, in the country of Honduras. This University has a special peculiarity: Its students are obligated to live, study, and work on campus and agree to very restrictive rules. Learning is guided by the "learn by doing" principle, which requires students to work in groups of ten in different small business enterprises of the University; I was always the only girl in the work groups because of the male: female ratio of 5:1. Many other girls told me how lucky I was and I agreed with them at first. Later though, after I had been forced to confront so many bugs throughout the year, I changed my mind.

In addition to all the predictable chauvinist characteristics of the guys from my group, Zamorano had its infamous recruit intern system. Basically this was about all seniors, juniors and sophomores making life impossible for lovely and sometimes lonely freshman people. Sadly this was the situation I found myself in – a freshman, and a female!

Anyway, during my first days at Zamorano, male coworkers and seniors began throwing my body into every single pond, river, and anything else that contained water. As a result I soon ran out of clean and dry uniforms. All these wet clothes were stuffed into my dirty-clothes bag inside my closet, day after day until cleaning day came.

To make things even worse, my dorm hall had yellow lights. I now understand (I didn't then) why so many May bugs (we called them "ronrones" in Spanish) were always crawling around my room and getting in my hair after I returned wet and tired from work.

It was just terrible, having no warm clothes, ronrones all over the place, and added to that the cicadas making that awful noise outside. I was barely able to sleep.

Toward the end of that long first week my roommate began complaining intensively, about getting bitten while asleep. Of course, I did not pay attention to her complaining because my own life was so miserable at that point, dealing (as I was) with the guys, recruitment (the intern discipline system of Zamorano), and classes! Life was definitely not good.

Finally, the end of that awful week arrived and more importantly, the laundry day. By then I was more than ready to trade my wet uniforms in for new ones. As I walked to the laundry service with my bag bouncing against my back, I kept

Illustration by D. Castillo

noticing peculiar bites and tickles. I had no clue what the cause was until I handed my bag to the laundry people. When they dumped out the contents there were hundreds and hundreds of ants inside my dirty wet clothes bag, crawling frantically from my pants to my shirts! I couldn't believe my eyes. I naturally did what any red-blooded girl would do: I dropped the bag and ran, screaming back to my room.

My roommate was there, still feeling blue. I told her," I totally and completely understand your complaints. It's my entire fault. I am an insect magnet!"

Fishing in the Entomological Stream

I had hoped the laundry people would not notice, but after colonies of ants arrived at the laundry room week after week, I was disappointed to get a notice from them. Their urgent note read: "We don't need more bugs in here, we have enough in Zamorano already" "Yeah you are telling me that. I couldn't agree more."

After long, continuous and numerous complaints about my ant-infested clothes, bug bites, insect noises, as well as a plague of ronrones; my roommate and I arrived at a decision, we were simply going to find a way to live with insects! But how?

To begin with, I gave up on trying to get rid of the ant colony every weekend; instead I set them up a little home just outside my room: one pair of wet pants every two days until they dried out. For the ronrones, I simply turned off all the hall lights during the night, thereby allowing them to migrate to other dorms. As regards the cicadas, sadly we could not kill them all; but, in fact, we came to miss their droning songs when their season had passed.

So I must admit my first experience with insects helped me to learn that they are more than something that just bugs you. I came to see that it is incredible how many different types of habitats, and niches they inhabit. And I slowly came to be glad for all the benefits they give us, sometimes even when we don't see them.

The moral of my story is:

Remember: Things you hated in your past may become your passion in the future.

ABOUT THE AUTHOR

After a large dispute over what to call the new baby, her parents chose the name Diana. She is the youngest of three children born to the Castillo Family on the south coast of Guatemala. The apple of her daddy's eye, it wasn't until she started high school when her adventurous spirit was unleashed. She wanted to know more about the world and questioned what could be found beyond Guatemalan borders. She left home to attend a prestigious, but male-dominated college in Honduras, where she studied agriculture. Because of the University location, Diana traveled all over Central America, the Caribbean and parts of Mexico. Like many girls her age, she experienced crazy adventures and even received the nickname "Party Animal" while attending college. Upon graduation, she decided to settle down (a little bit) and get a job in the United States in a Biocontrol lab at Purdue University. Through her hard work in the soybean field she was accepted to pursue a Master's degree in Entomology. Good meals, fine wine, and stimulating conversation top the list of things that Diana enjoys. Diana loves meeting people and experiencing new things, the reason traveling remains her passion. Diana dislikes dirt, loud eaters, and gets nervous when people cut their fingernails in her presence.

"The greatest motivational act one person can do for another is to listen."
-Roy E Moody

Essays on Insects and Life by Students of Both

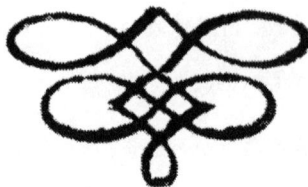

AUSABLE RIVER

By Carolyn Foley

n a cold September night in 2002, I met a great love. The day had started too early for my liking. The night owl in me wasn't keen on meeting my aquatic ecology classmates behind the University of Windsor's Biology Building at 7:30 a.m. My eyes still heavy with sleep, I climbed groggily into a van and settled in for the four hour trip to Arkona, Ontario: land of streams and fossils. We followed strings of rural roads and passed through a handful of backwater towns before finally arriving. Our purpose was simple: experience the ins and outs of real, live stream ecology. The next two days would be crammed full of carrying equipment up and down 100 steps of stairs, sleeping on the ground, and setting up experiments. I was thrilled.

We conducted experiments on two waterways: Hobbes Creek and the Ausable River. For each of these, we measured mean width and depth of the channel, calculated water velocity, put algae-covered rocks into Plexiglas chambers designed to measure carbon uptake, stunned fish to the surface of the water to identify the species present, and collected macroinvertebrates via kick-sampling. While most of the experiments were mildly interesting, it was kick-sampling that began to turn my head. This highly technical process involved dancing in the water upstream of a net, being careful to stay on your feet despite the rushing current and slimy rocks. After 60 seconds, the net was emptied into a shallow pan. Water pennies, stonefly nymphs, and other aquatic invertebrates that had been dislodged by the dancing were collected out of the pan and preserved for their ride back to Windsor.

I was quickly recognized as the best kick-sampler in our group, i.e., I was the only one willing to look like an idiot in front of the professors. There were a few close calls where I almost wound up face-first in the water; our teaching assistant, recently returned from another task, stood ready with his camera, hoping to immortalize my fall on film. To his chagrin, I stayed upright and dry and, grinning, asked where he had been. He replied that he had been setting up another experiment. This one involved drift nets - fine mesh nets with jars attached, placed in the water and meant to collect whatever happened to float by. The contents were to be removed and preserved in ethanol every half hour for 24 hours; the samples would be sorted back in Windsor. Each member of the class had to man the experiment for four hours, and I had drawn the 12-2 shift: 12-2 a.m. Sunday morning, then again 12-2 p.m.

Sunday afternoon. This time slot was infinitely more appealing to me than beginning our trek at 7:30 a.m. had been, though I was a bit wary about wading around in a river in the middle of the night. I didn't have much time to contemplate my lot, though: there was too much work to do, and so I continued measuring, calculating, and forcing invertebrates out of their cracks and crevices and into our net.

At 6 p.m. on Saturday, we went back to camp for a satisfying chilli dinner. At 6:30 p.m., I helped light a fire that was still burning when I emerged from my tent five hours later. This was the start of my drift net shift. I turned on a small flashlight and walked carefully down the stairs that led to the streams. Once there, I was treated to hot chocolate made on a Coleman stove, then told to come along and learn my job for the next two hours. I pulled on a pair of rubber chest waders and followed the professor into the darkness. He showed me the way to the nets, how to empty them, and how to properly label and preserve the invertebrates. He then bade me goodnight and headed back to camp. At the collection time half an hour later, I was on my own.

At midnight I walked through the woods toward the river, still wearing my trusty waders. The moment that I left the shelter of the trees, that river became mine. The moonlight shone back at me from the water, making it easy to see where I was headed. I stepped cautiously, avoiding the slippery rocks and quick-moving water: if I hadn't crashed and burned while kick-sampling, I sure didn't want to do so now. As I looked down at the rocky bottom, I was reminded of my days as a cross-country runner. During races through the woods, I would constantly scan the forest floor for stumps that might trip me. Looking down seemed a good idea at the time, but it made me miss the beauty of the trees, the leaves, and the occasional squirrel or deer in my path.

Illustration by J. Green

Recalling this, I stopped and forced myself to look up. In a few moments I was able to drink it all in: white lights playing on rippling waves, rocks strewn onshore, trickling water droplets chasing each other through riffles, slower droplets sitting stagnant in pools, and me, tramping through the stream in my clunky waders. I sighed: my river. It was beautiful. I did not belong in this magical place, but I was glad to be a part of it. I bent over the drift nets and massaged the sample toward the collection jar, as I'd been shown. Further joys of the collection would be revealed weeks later in the lab, in the hours spent searching through leaves and algae for larval midges and dragonfly nymphs. At that moment, though, all that existed were the invisible invertebrates floating down the river and I, each of us silently making our way toward dawn.

My replacement showed up at 1:30 a.m., and I was sorry to share the river with another human. I left her at 2:00 a.m., and climbed the 100 steps back up to camp. As I poured water on the still smouldering ashes of the fire, I contemplated what had just happened and came up with one thought: "Marvellous." That night, among the rocks and currents of the Ausable River, my love affair with field ecology and invertebrates officially began.

ABOUT THE AUTHOR

Carolyn Foley hails from Amherstburg, Ontario, Canada, and has spent most of her life trying to figure out what to do next. At times, she has wanted to be a ballerina, a nurse, a math teacher, an English teacher, an Olympic athlete, a stunt pilot, an astronaut, a marine biologist, an engineer, and a professional traveler. A love of the outdoors led her to pursue graduate studies in landscape ecology and entomology, and her enthusiasm for the subject is infectious. She enjoys working with the public and hopes that she can convince people to care about the earth that they are slowly destroying.

I AM AHAB

By M. Walter Baldauf

osta Rica: The Airport

Stepping off of the plane in Costa Rica I experienced climate change for the first time in my life. Yes, I had been to Canada, England, and France, but I had never been close enough to the equator to really feel a change in latitude. My aim in Costa Rica was to study biodiversity and coffee production as the final course of my undergraduate college career. Commencement had taken place the week before; after this trip I was off to graduate school. I was not on the trip to expand my insect collection, with iridescent blue morphos as big as birds or bright flashy tiger beetles. Forget them. I was interested in only one insect, a beetle, and nothing else.

The States: Briefing

I had learned about this beetle back in the States before leaving for Costa Rica. My undergraduate professor, Dr. Oseto, had told me of these large insects during an insect taxonomy lab. Previously a student in Costa Rica had heard a knocking outside his hotel room. He opened the door to discover three golf ball-sized beetles banging against a light fixture. They were nocturnal. This was the key thing that I needed to know; I was determined to catch this Moby Dick of an insect.

Costa Rica: Traveling

I had two weeks to find this beetle. I assumed it would be simple, since insects were everywhere in Costa Rica. The air was so thick with them in the jungle that I had to revert to long sleeves and pants despite the boiling temperatures and high humidity. This was nothing like the Indiana heat I had known in my youth: that heat was escapable with air conditioning. This was Costa Rica. Full clothing in steaming heat just had to be accepted, even welcomed. The alternative was exposing my flesh to blood sucking flies.

Every night for a week I waited patiently for the sun to fall behind the mountains and the night to spring to life. I peered around the lights of the hotel long after most of my colleagues had succumbed to acute inebriation. Sleep deprivation was of no concern. I figured that the lack of sleep could be made up when I got back to the States, and besides all the humidity made the sheets feel like a wet nap. It was the end of the week and there was no sign of the beetle. Other students spent their days collecting butterflies and watching birds. They had found their white whales in daylight, in large numbers, while my leviathan was swimming through the darkness of night, out of sight, perhaps never to be discovered.

Costa Rica: National Park

There was a man standing next to a slide projector talking about bats. Costa Rica is home to many of the different bat species of the world, and the reserve where we were specialized in bat research. The long, hot lecture ended with a tour of the collecting nets they set up every night. I had never seen a bat still in the wild, and that night they had captured two. One was pregnant and was released, but the other was bagged to be taken back to the reserve. That was not however, all they had caught that night. My wait was over; the sought-after beetle was in sight, finally I had the beast in my grasp. I worked feverishly to free it from the bat net, but it had tangled each of its legs. A spider truly could not have wrapped it tighter. I almost had it free when the guide called everyone to follow him back to the hotel at once! I had to leave. I had to leave, the beetle had won our struggle and so it escaped me. My white whale had slipped away and gone back into the darkness and the depths. I did not stay up collecting that night. There were no lights outside and I knew the beetle was still out there entangled in the net. I realized that it had purposefully tied itself up in knots to tease me and frustrate me. It had probably been caught in the net before and knew it would be set free in the morning when the nets were taken down at dawn. It was taunting me. That glimpse of what I could have had in my hands was worse than never seeing it at all.

Costa Rica: Cahuita

The days passed. They were spent trudging through the forest listening to other students "ooh" and "aah" over the treasures they had found. The nightly watches continued. Every evening I would sit and wait for the beetle. It never came. The last night we were to stay in the forest was in the hamlet of Cahuita. This was a small village located on the most eastern and southern tip of Costa Rica, just a stone's throw from Panama. Cahuita is the home of two national parks, white sand beaches, and sky-blue water. I saw a sloth and monkeys here for the first time in the wild, but that is not what I was searching for. This was my last chance to catch the whale.

Costa Rica: The Last Stand

That morning was spent touring a national park, followed by an afternoon at the beach. I was swimming in the ocean, enjoying being tossed around by the waves. But at one point I was briefly pinned to the sandy floor: this seemed to be a warning, so I took heed of it and dared not venture farther out. I swam to shore, headed back to the hotel, ate dinner, and sat by the pool. I had an Imperial, the local's choice of beer, and reflected on my trip. I thought about the wonderful things I had seen and wondered how the last night of my hunt would go. Would the whale continue to elude me, or would I, like Captain Ahab, join the ranks of those who had managed to catch it. As darkness fell, something happened that I will never forget. I heard the clumsy flutter of heavy wings and felt the whiz of a golf ball graze my chest. It startled me, and for a moment I did not know what it was. I heard the sound again, but this time something landed on me. I grabbed it in my hand and held it under the light: there was my white whale! The beetle was twice the size of any other beetle I had ever seen. Its body was chocolaty brown with ebony legs. It's teasing was through: after relentlessly searching night after night it had found me.

I quickly ran to my room, and couldn't help but notice how powerful the insect was. I could distinctly feel the push and pull of each leg as it tried to escape from my fingers. I had always been taught that insects are extremely strong for their size, and with this monster in my hand I could truly appreciate it. I searched for a container to store my treasure in for the trip home. I found a liter water bottle and put the beetle in. Now all I needed was alcohol to preserve it. As I rifled through my pack I could hear it. The power in those scrapes and scratches actually - I hate to admit it - scratched at my heart and stopped my search. I never found the alcohol; instead I observed the beetle for several minutes as it methodically scratched its way around the bottle, searching in a vain looking for a way out. I had pinned dozens of insects for my collection, but I knew I couldn't pin this one. Taking this specimen back wouldn't do it any justice. It had no distinct markings. It was large, but its true beauty was in its power, and in the way it moved. There was no way to pin that. I now understand why people watch birds. The fascination is not something that is tangible; it is in the experience, in the moment, and in the memory. Our paths had freely crossed. The beetle put its fate in my hands, my hunt was over and the ending was mine to make and I made it.

I put it back out to sea.

ABOUT THE AUTHOR

Walter is 23 years old and on track for living a long life following advice from advertisements and suggestions of health professionals. He feels that long life is not what's important, though, because, as the quote says, life might be over tomorrow. He'd rather live a full life and appreciate everyone and everything around him, even if he can't stand some of them. During his time on this planet, he has come to realize that the more he can teach people the longer he will really live. Fame, fortune and immortality from individual pursuits are unlikely, but teaching someone to tie their shoes will continue on long after he is gone. If the lesson about tying shoes is the only thing that lives on after he's gone, he's okay with that, because he will have helped at least one person not trip and fall on their face. Walter's best friend, always heavily medicated with pills and rock and roll, shared this quote with him:

"Some days the guns loaded and some days the guns not, but regardless it's always in my mouth."

- J. W. Furniss

A SUMMER JOB

By Marissa McDonough

I t was a balmy July morning in Indiana; one of those stagnant days with no wind to cut the heat. The birds sat listlessly in the trees waiting for a cooler time to sing. The sun, out in full, poured more heat into the day. In the forest, however, the weather was more tempered. The sunlight fell to the forest floor in scattered intermittent rays. The temperature was cooler; a little more endurable. Leaves littered the ground, remnants from the year before. Life sprang up everywhere: new trees striving to outrace one another and meet at the top of the canopy; grass inching up through the fallen leaves; forest flowers were alive with vibrant purple petals.

In the midst of all this life was the distinct sight of death. Rotting flesh is a sight not easily forgotten. A former vessel of life, a corpse is like a curious fountain of resources from which many organisms drink. A dead body lay on the forest floor.

A van drove into the woods from the clearing and stopped near the dead animal. The research team, consisting of my boss, my friend and me, got out with a news reporter quick at our heels. Everyone except the reporter was dressed in shorts and t-shirts. We were spattered with dirt because we had been hard at work earlier that morning in the field adjacent to the forest. The reporter carried his pad and pencil as well as a camera. He was clad in freshly pressed jeans and a polo shirt. The team, followed by the reporter who chatted merrily, trekked deeper into the woods to the site where the lifeless body lay; the reporter got very quiet. It was only when they got very close that they could tell this animal was formerly an adult pig. As the scene came into view, the reporter paled and leaned on my arm.

"Whew. I thought I was going to lose it there for a moment," he said as he bobbed queasily.

When I saw he had recovered, I began showing him around the site and explaining exactly what a forensic entomologist does. Even as a summer job, this was the best time I had ever had. I liked the research and enjoyed getting dirty. It was fun working in the dirt and collecting bugs off of dead pigs, something I did everyday. It was also fun to bring new people to the research site and share my knowledge of this fascinating research, which my mentor was nice enough to introduce to me.

The carcass that formerly enclosed the weight of a full-sized pig sizzled with little white maggots. There were thousands upon thousands of them fighting for food and for life. Life is a funny thing. It is weird to think about how simple but also complex it can be. This corpse, formerly a farm animal, was now reduced to a blob of decaying organic matter. What once had contained life, had now been transformed into a resource to create life for other organisms. Flies were dive bombing our heads; trying to find a piece of real estate to release their eggs in hopes to continue their cycle of life.

I explained to the reporter that the eggs hatch into maggots -- those little wormlike creatures that evoke the gag reflex when seen by the normal public. All these little guys do is eat. It is their only goal in life. Eating means that life will continue, even when their food is putrid tissue. I snickered to myself when the reporter turned green as I scooped a pile of maggots off the carcass with a gloved hand to demonstrate the sheer number of them and prove to him the heat they generated. I confess I found it fun to gross people out. The sound of the camera shutter filled my ears as the reporter took pictures of the writhing creatures dripping from my hands.

"How do you deal with it? The smell, the gore, the reality of it all," he asked in wonder as he pulled his shirt over his nose.

A giggle erupted from deep in my gut. It was so much fun to bring people here who thought it was no big deal; just like the television show but not exactly. "You can't *smell* what is on the television." The most common question from people who don't commonly work with these things was the one that the reporter just asked.

"Someone has to do it," I replied with a smirk.

It was in that moment that I recognized that what I wanted to do with my life was to be a forensic entomologist. I realized I could be a voice speaking for those who could no longer speak for themselves.

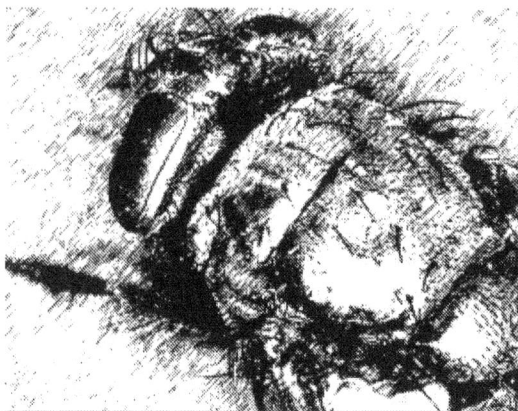

A blowfly

ABOUT THE AUTHOR

Marissa McDonough is the older of two sisters who were raised in Vernon, New Jersey. Marissa received her B.S. in Health Sciences at Purdue University before moving over to the Entomology Department to receive her M.S. in Forensic Entomology. Marissa is married to Marcus McDonough and they live in Lafayette, Indiana with their four cats, dog, piranha, and feeder fish that continue to reproduce because the piranha is too blind to see them to hunt and eat them. Marissa enjoys running outside, traveling, and eating lunch at the Irish Pub. Marissa is a germaphobe who washes her hands non-stop, and avoids sick people, but she has a soft spot for animals and would take in every stray if she could.

"Necessita c'induce, e non diletto."

(It is necessity and not pleasure that compels us)

- Dante, Inferno (XII, 87)

STICKS AND TOADS

By Philip K. Morton

I once listened to a seminar by Bobby Corrigan that began with a quote from E.O. Wilson. The quote went something like, "To the lazy hunter, the forest is empty." Wilson is famous for his work on ants and for applying the principles of science to other aspects of life. Although Bobby Corrigan currently works on rats, he is also a trained entomologist. I believe that both of these men view the world in a way that enables them to see things that others would walk obliviously past. I strive to be like Wilson and Corrigan by acquiring the ability to see things differently than other people.

As a boy, I collected anything that crawled, slithered, flew, hopped, scampered, or died - much to my mother's distaste. She almost always made me get rid of whatever I brought home. This was certainly the case when I dragged home a dead opossum by the tail, my eager boy's mind filled with the notion of watching the succession of decomposition in my backyard, then reassembling the skeleton in my room. Understandably, Momma did not want any part of this rotting corpse in (or around) her house. Being a dutiful son, I gave up on this idea easily enough. Nevertheless, I did try hard to talk Momma into allowing me to keep any *living* thing that I might get my hands on. On a few occasions I was victorious. To my mother's credit, she ensured that we went to the zoo or the science museum at least once a month. I believe she did this to satisfy my hunger for repulsive biological things so I would not bring them home to her kitchen. Thanks to my peculiar enthusiasm for living things, I was pegged as the weird one in my family (for now, I will spare you the other stories and keep to the current one).

When I grew up and it came time for me to go to college, I did not know what I wanted to study. I enjoyed science and math, so I thought some sort of engineering degree might be the right thing for me. When my top choice school did not offer as much scholarship money as I had hoped, I explored other options in both professional specialty and institution. This was the point when my father told me that he knew someone who could get me a scholarship if I was willing to study entomology. Immediately, (I'm embarrassed to admit) I asked, "What's entomology?" After Dad explained it, I said, "Yes! That's *exactly* what I want to do!" I could not believe that someone could actually get a degree studying insects and, furthermore, that I had never even heard about it. Upon learning of this incredible possibility, I quickly made my decision, and soon was officially on my way to becoming an entomologist – with my parents blessing, to boot – life was good.

As an entomology undergraduate student, I took a taxonomy class in which the majority of my grade depended on an insect collection. At first, I figured this class would be easy because all I had to do was go out and pick up a few insects in order to get an 'A'. On one of our first class collecting trips to a small pond, however, I began to change my mind. Our professor declared that it was a great place to collect aquatic insects. Naturally, I jumped right into the mud on the banks and started looking. I searched and studied the banks of that pond for any sign of insect life, but could not find anything. Then a graduate student walked right up beside me, bent over, and effortlessly plucked a water scorpion out of the water with forceps. It was as if that water scorpion had been holding up a giant sign (invisible to me), that read, "Here I am!" I took another look, and to my surprise there were water scorpions all over the pond, big ones too, about as long as your finger and as thick as a twig. After this, I was rather embarrassed that I had not been able to see them earlier, but I did not tell anyone. Since I could see them now, I thought I had unlocked the gate and now I could find *any* insect.

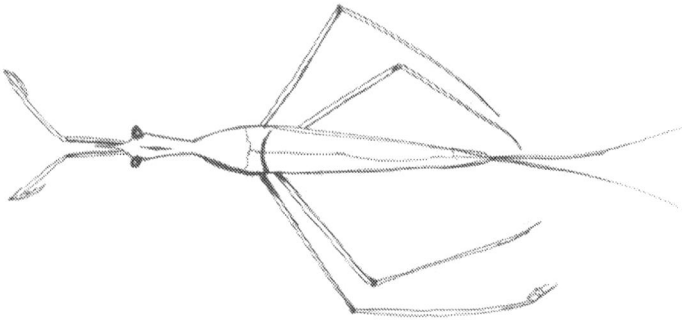

A water scorpion

On our next collecting trip we visited a stream. It was narrow, yet moving with a deliberate pace. One of my buddies immediately bent down and scooped up something into his collecting jar. Then he leaned over to me and said, "There are toad bugs here." I was excited, even though I had no idea what a toad bug was. A little later on, I was walking up the stream and as I straddled it with both feet on the bank, stuck in the kind of mud that appears solid yet gives easily under your weight, I noticed tiny frogs hopping away from my feet. When I took a closer look, I could not believe my eyes! These were the toad bugs, dozens of them! Hopping everywhere, these little insects had identical coloring and movements like miniature frogs. I even accidentally caught a small frog because it moved just like the toad bugs -- or the bugs moved like the frogs -- whichever way you prefer. This was astonishing, and I could not wait to collect more.

Thanks to these collecting trips I began to improve my observation skills, and started looking more carefully at the environments where I collected. Soon insect collecting became more fascinating to me. I became steadily better at finding insects, particularly when other people were collecting insects close by, because they helped to keep me focused. In time, I came to believe I had an excellent eye for spotting insects. So, when I went fishing with my Uncle Michael, I took my collecting gear essentials, such as my trusty net and freshly-charged killing jar, and planned to collect insects as well as fish. By that time I was 'the' entomologist from a university, delighted at the prospect of impressing my uncle with my entomological prowess. To my dismay, I didn't find much except a single pleasing fungus beetle which, regrettably, slipped through my fingers. I also unintentionally found several colonies of fire ants while I was fishing. Although I had brought my net, killing jar, extra vials, syringe, ethanol, and anything else I might need to collect a prized specimen, I did not think to dress appropriately, and thus had on flip flops and shorts. As I cast out my line, I felt slight movements across the top of my feet and between my toes. The intensity of the tingling movement increased, and I felt that a glance away from my pole was warranted. Just as I looked down the fire ants let loose their formic acid and thousands of tiny, burning stings erupted all over my feet and legs. Uncle Michael was amused by my bug dancing that followed.

Fishing in the Entomological Stream

I was set back mentally because I was not able to find insects without the influence of other collectors nearby. This led me to realize something of great importance, namely, that there is an unseen world that thrives all around us. That world can have a great impact on our everyday world, and those who see both worlds simultaneously have a great insight into the true nature of life. The unseen world, for me, is the world of insects. They are small creatures that are evolutionarily simple, yet at the same time, infinitely complex. While we continually learn more and more about their secret world, I suspect that it will be long after I am dead, if ever, before we are truly experts on their simple life forms.

Currently, I am more aware of everything around me because I attempt to notice the little insects that crawl around me every day. Notice that I include the word 'more' before the word 'aware' because I continually miss things, or see them only after others do. I am becoming better at seeing things clearly though, and occasionally I see them first. I make a point about seeing them first because I feel that the person who sees something first is inherently aware and always has open eyes. This is the hunter who is not lazy, who sees that the forest is teeming with wild things nestled within the vastness of its finite edges. This hunter will set new limits for the world.

ABOUT THE AUTHOR

Philip K. Morton was born in Liberal, KS, to Derrick and Pamela. He has one younger brother, Paul. He is happily married to Ashleigh Morton and they currently have one beautiful daughter, Violet, his greatest accomplishment to date. Although Philip was born in Kansas, he lived most of his life in Oklahoma where the majority of his family still lives. He received a Bachelors degree from Oklahoma State University in Entomology, with an 'A' in his Systematics class, and is currently pursuing a doctorate degree in Entomology at Purdue University in West Lafayette, Indiana. As an undergraduate, Philip worked on stored product insect pests and phylogenetic relationships of South American mouse opossums. In his doctoral study, he works on the population genetics of the Hessian fly. His hopes for the future are that he will give Violet (and any of her possible siblings) a fun and fulfilling childhood full of adventures, fishing, and insect collecting, while equipping them for their future.

"With great power comes great responsibility."

−Uncle Ben, Spiderman

THE DANGERS OF TREASURE HUNTING

By Ashley Walter

nce, when I was shopping with a friend, he teased me and said that shopping with me is like following a magpie - I'm attracted to anything shiny. I don't deny it. I like things that sparkle. That's probably why geodes have always held a special appeal for me. The outside resembles nothing so much as a petrified cauliflower, but break one open and you will find beautiful sparkly treasure. I spent countless hours as a child combing the streams around my house in Danville, IL looking for geodes. Once I got to college I gradually forgot about my earlier obsession.

Indiana was not my first choice as a place to finish up college but unfortunately for me, it made financial sense. I had taken a couple of years off school to bum around and figure out what I wanted to do with my life. I learned two things from this scholastic break: 1. I am horribly irresponsible with money, and 2. I don't want to wait tables for the rest of my life. Totally broke and with little good grace, I succumbed to the wishes of my family and applied as a transfer student to Purdue. I was accepted into the Entomology Department and once there finished up my bachelors degree. In my senior year I was offered a position in a masters program. My project would be to develop an educational program to help protect forests from invasive species. Since I was still sort of at loose ends, I figured "What the hell?" and accepted.

I pouted for the first two years I lived in West Lafayette. I liked my advisor, my classmates, my project, my apartment, etc., but I could never fully reconcile myself to living in Indiana. As a native of Illinois, I had always looked on Hoosiers with contempt. Also, I had long ago promised myself that when I left Illinois I would move somewhere closer to the ocean. Indiana was just not part of my master plan. It wasn't until a trip to a park in Brown County that I discovered something I could really love about this state. According to one of the displays in the park's nature center, the southern half of Indiana is part of something called the "geode belt". At the time, I had no idea what this meant but, liking the sound of it, I filed this tidbit away for future reference.

Later on that year, I went on an insect collecting trip with a group of fellow graduate students. I have to confess that I had no intentions of collecting any insects. Don't get me wrong, I think they're fascinating creatures and I do try to observe them when I'm out traipsing about in the wilderness. I just have no

Miscellaneous geodes

desire to have a collection of dead ones. If nothing else, a bunch of dead insects wouldn't match my apartment's décor at all. We camped outside of Bedford, Indiana, in the southern third of the state. I suspected this might be part of the rumored geode belt and I had every intention of conning some of my fellow campers into finding a creek so I could verify the legend.

Three of us arrived before the others. Having nothing better to do, we decided to wander down an overgrown path through the woods to see what was out there. I agreed, hoping this might be my chance. After fighting our way through weeds and some sort of horrible bush with razor-like thorns, we found a little limestone-bed creek. We scrambled down the creek for about 10 minutes and I didn't see a single geode or any other rock of interest. I felt letdown but tried not to show it. "Some 'belt' this is," I muttered to myself. We did make acquaintance with an alarming assortment of stinging and biting insects. Accordingly, we eventually decided to turn back and head into town for some supplies. Beer rations were low, I wanted s'mores materials, and I had seen a creek running through the eastern part of the town on a map that I had studied on the drive down. I figured if I could get my friends to drive into town for food, I might be able to lure them to this creek.

I told my two companions about the geode belt I had read about and that we could potentially find hundreds of them just laying around. I imagined the creek to look like a particularly rustic shoot for a Tiffany's ad campaign- glittery treasures everywhere. My friends graciously agreed to drive around and find the creek shown on the map. We decided we should go hunting before replenishing said supplies so there would still be plenty of light to pick up geodes by. After much driving back and forth we found a small park with a shallow river running along its western edge. I put on my river sandals and worked my way down the bank and into the stream. There sure were a lot of rocks strewn about the bed of the stream. Suddenly, with a shout of joy and amazement, I realized that nearly every rock I could see was a geode! Of every conceivable size and shape! I had never in my life dreamed that such a place existed. I quickly showed my two companions what to look for and how to break the geodes open by whacking them between two rocks to see what's inside. They took to this primitive and strangely satisfying activity quickly.

It was fantastic. Before long, we started making piles of geodes to haul up to our truck. We splashed our way up and down the creek, combing it for the biggest geodes we could find. Our piles grew to ridiculous proportions. We had to use the beer cooler to carry our treasures up the bank to the truck. Though enthusiasm was high, after a couple of trips even I reluctantly had to acknowledge it was starting to get dark. We decided to stop for supplies at the Wal-Mart and grocery store we had seen on the way into town. I wanted to buy a pair of shorts to facilitate wading in creeks; I needed some plastic tubs to hold all of my geode booty, too.

My friend Alana and I went into Wal-Mart while our other companion, Paul, headed to the grocery store. I immediately searched out the clothes section. I grabbed a couple of cheap pairs of running shorts and went to try them on. I headed over to the fitting room and was handed a tag from the sweet little old lady in charge of monitoring it, then went inside. Alana had gone to find the plastic tubs and planned to meet me outside the fitting room momentarily.

I went inside the tiny cubicle and locked the door. As I was taking off my pants, I noticed something that looked like a mole. Upon closer inspection I realized the unfamiliar mole was actually a tick. A tick with its head buried in my hip! Now for all of my entomological aspirations and training, and for all the time I've spent wandering around in the great outdoors, I have to admit something: I really, really hate ticks. Truly, their very existence is offensive to me and finding one on my person was alarming. I confess I completely and totally lost my cool.

I immediately started slapping at the thing in a complete panic, while a stream of curse words that would make a convict blush spewed from my mouth. I was still half in and half out of my pants so when the tick was finally dislodged by my flailing at it, I tried to hop backwards away from where I imagined it had been flung. In the process, I lost my balance and crashed into the door of the dressing room. Just then I spotted another suspicious looking spot on my other hip. A quick examination confirmed it was a second tick. The violence of my curse words rose in intensity as I began slapping frantically at that one.

At this point I realized I was in the grip of panic and so tried to regain some composure. I told myself that I am an entomologist who is working towards saving the forests where these creatures live and that I should not be afraid of something as silly as a little tick. With that in mind I managed to get my pants off. I took a few deep breaths and picked up a pair of shorts. While slipping them on, I noticed a leaf stuck to my foot and reached down to flick it off. When I did so, it twitched. It dawned on me that the leaf was actually a leech and the panic resurged with greater urgency. So did the swear words, which are embarrassing to remember. Ticks, at least, were familiar if horrifying territory. I had never seen a leech in real life before. That was the last straw.

I flung the leech such that it hit the wall of the dressing room with a splat and stuck there, wriggling. I ripped off the shorts, threw on my pants, and stumbled out of the dressing room. I had completely forgotten about the kindly old lady waiting outside. I was completely disheveled, hyperventilating, and still muttering a bit under my breath. I suddenly realized that this poor woman had heard me cursing and banging around inside the dressing room. With a look of concern on her face, she asked me if everything was all right. I managed to compose myself enough to say that everything was indeed fine and that this particular pair of shorts was exactly what I was looking for.

In the meantime, Alana had found the plastic tubs and was waiting for me to come out of the dressing room. When I came out I was pale, shaking, and looking completely distraught. She looked alarmed and asked me if everything was OK. I looked back at her, truly shaken, and managed to gasp out "two ticks...... and a leech." She was taken aback for a moment, but then burst into hysterical laughter. She kept laughing as we moved towards the store exit. Tears came to her eyes and she had trouble steering the shopping cart. I was a bit nonplussed at her display of sympathy and expressed my desire to get out of this horrible store and have a cigarette to calm my rattled nerves.

We made it to the truck, her laughing all the way, and sat down to wait for Paul. My hand was trembling so much I had trouble lighting my hard-earned cigarette. After calming down a bit, I speculated on what would likely happen to the next unsuspecting customer who used that dressing room. Two hungry ticks were running around looking for a new victim and a leech was waiting on the wall. This sent my friend into further gales of uncontrollable laughter.

As I leaned, traumatized, against our vehicle, I realized that I am a lousy entomologist. I like the idea of the outdoors with insects and little forest creatures but when it comes right down to it, I don't want to touch them. I want to save the forests but I want to do that from the safety and comfort of an office somewhere, preferably in a large metropolitan area. The reality of forests is often dirty and completely disgusting, as I had just discovered. I then looked into the back of the car and saw the mountains of geodes, just waiting to be cracked open. Eyeing the geodes, I considered that perhaps I was being overly dramatic. Perhaps there are some things in the natural world worth seeking, despite the occasional inherent discomforts. And yes, geodes are well worth the risk. But in the future, I shall never step foot outdoors without first bathing myself in DEET.

ABOUT THE AUTHOR

Ashley Walter spent a pleasant childhood in Danville, IL, dragging home rocks and piling them in her closet. When she was 12 years old, she moved to Champaign-Urbana and went to high school there. The rocks were replaced by clothes. She then started college at the University of Illinois, where she went for 3 years and then decided she needed a break from school. As she could no longer afford clothes, she started collecting rocks again. That collection was eventually dragged to West Lafayette, IN, when she transferred there to finish her undergraduate work. She lives there still and is attending graduate school at Purdue University. She's pretty sure she can never move out of her apartment because her rock collection has gotten completely out of her control.

"Anyone who lives within their means suffers from a lack of imagination."

-Oscar Wilde

HINES EMERALD

By Larry Murdock

I t is noon on an August day in southern Indiana and I am in a far woods, waiting and watching. The nearest road is half a mile away, the nearest human still further. I am sitting on a large stone by a little woodland pool. It is silent here except for the natural sounds of the woods, the muted rustlings and faint peeps and whispers of things unseen among the trees. It must have been exactly like this, here, in this place, two hundred years ago, and two thousand, too.

A seep of water from a stone outcropping a hundred feet up the hillside on my right becomes a trickle that drips downwards and pools on a stony flat among lichen-mottled rocks. That pool is at about my eye level. From there the gathering water glides over a mossy stone down to a larger pool, forty or fifty feet across, just below my feet. Feathery green ferns dominate the deep shade under the rock outcropping.

I am waiting on a sandstone boulder at the place where the streamlet from above enters the little pond in front of me. I am alert, attentive. Surrounding me in the broader sweep of things

— as the soaring red-tailed hawk above me sees it — is a horseshoe of rising wooded slopes. On the far side of the mirror of water before me (there is no wind at all, not a breath of it) are a few clumps of cattails wading in the shallows, eking out their lives in their element of water, soaking up the fractured sunlight that filters down into this extraordinary place.

On the rising ground around me there are big, old trees. I see not the slightest sign of humankind. I know that a century ago men with axes and saws and teams of mules came here to harvest the virgin timber. No trace of their long-ago visit remains; the treetops they lopped off have moldered away, the stumps they left have rotted, the road along which they snaked away the logs with their straining, snorting teams have healed again, reclaimed by the woods. The trees around me are mainly sugar maples and white oaks. A hundred feet behind my back is one particularly striking tree, a barrel-thick grey-skinned American beech with broken-off main limbs, its top blown out by a windstorm many years ago. Still almost sixty feet tall, it has only a few desperate sprigs of green leaves on one side. This forest behemoth is barely clinging to life, like an old man hanging on the edge of a precipice.

Soon that giant beech tree will catapult me into a state of awe.

Fishing in the Entomological Stream

The air is moist to saturation. I sweat in the steamy heat and the beads drop away, not evaporating, not cooling. The floor of the woods is brown with the random scatter of last year's fallen leaves. The rock on which I perch is hard without mercy. Ten thousand years ago it cracked loose from the outcropping above, shoved out by the inexorable freeze and thaw of numberless winters. One day, drawn by the inexorable pull of gravity, it came thundering down the hillside with a tumble and a crash. Ever since, it has been waiting here, this rock on which I squat, all those years upon years. Before my squinting eyes is a solitary shaft of sunlight shining on the water down a channel through the treetops. That patch of light is the only bright spot in this dark, haunted place; it dazzles amid the surrounding gloom like a lit match in a cave. I squint into the light. What I am watching for, what I have come here for, will appear in that shaft of light. I hope. If it appears at all, I will have to be lucky, extraordinarily lucky.

With the imperfection of memory, I recall what the book said: "It is found near woodland ponds and sluggish streams ... rare, and very old it is ... a life form that arose two hundred million years ago ... it has survived little changed ... this rare species may be extinct in Indiana, though it has been seen in Ohio, Illinois and Michigan ... prefers shaded woodland ponds ... hovers in patches of sunlight".

No place fits that description better than this, here, where I wait. It is the perfect habitat for my quest, the rare Hines Emerald. That's its name. If the Emerald is here, and if it is the right season, if the gods smile, I shall see it.

Minutes pass, then an hour. My thoughts wander. I wait and watch and time creeps on. I muse that when the stone on which I sit came thundering down the hill long ago, it may have startled a mastodon, or caused a great cave bear to turn its head. Maybe it frightened a family of deer – I can see them leaping and bounding away in terror.

These woods are haunted by spirits, I know, I can sense them around me, just out of sight. The Druids worshipped in their sacred groves, knew the magic of such places. Anyone who wanders far into the woods alone and sits quietly and lets himself simply be aware knows it, too.

The words of Walter de la Mare drift to mind:

> *Very old are the woods;*
> *And the buds that break*
> *Out of the brier's boughs,*
> *When March winds wake,*
> *So old with their beauty are --*

Fishing in the Entomological Stream

Oh, no man knows
Through what wild centuries
Roves back the rose.

On the mirror of water before me I watch water striders skating, their wispy weights barely dimpling the surface of the shallow pool. I see through the limpid water little oval lights moving on the shallow bottom. They are created by the microscopic dimples formed by the striders' feet. Those dimples are lenses that create far bigger lights on the bottom. Those lights skitter below the striders exactly as fast as they move above. "What is it like to be a water strider?" I wonder. I try to picture myself as one. All day long I skip and skate along the surface, pouncing on a fat aphid that plops into the water, leaping on any fellow strider that strays into my territory, zipping away when a diving beetle oars under me and threatens to rise and gobble me up. It is another world, miniature, yet as big as my own to them, and far more frightening.

"Floating on the silent surface of the pond, I almost cease to live and begin to be", wrote Henry David Thoreau more than a hundred years ago. When I read Thoreau in high school, those words seemed romantic merely, maybe even pretentious. Here, by my silver woodland pool, they take on a far deeper meaning, ring utterly true. By this woodland pond I understand Thoreau for the first time, here among the trees and the woodland water.

Two hours pass, then three. For me there is only listening and reverie. From time to time I shift my position to relieve the numbness in my legs and escape the discomfort of the stone. I slap at mosquitoes, too. I do so quietly, though, almost warily, wishing not to be irreverent, not to offend the wood spirits whose home I have invaded. I am like an ice fisherman in that respect, oblivious of time, or like a hunter who doesn't care whether his quarry ever appears. Such is the pleasure of hunting and fishing, peace and quiet in a place of contemplation and anticipation.

Time stands still.

In my reverie, I gradually become conscious of a faint skittering sound, a just-perceptible rattle. It floats lightly upon the windless silence of the woods. It causes me to think of microscopic reeds flexing and bending at high frequency. My drifting consciousness, at first not fully grasping what is making the sound, still confused, like a man waking from a dreamy sleep, slowly and slowly I become aware of the skitter, realize that it is coming from the patch of sunlight before me. In slow motion my mind at last begins to notice what my eyes are already seeing:

In the shaft of sunlight above the water is a large dragonfly, so close it looks two feet across (though its wingspan is no more than four inches). It hovers there and watches me with its huge glittering eyes, green eyes, an emerald green glowing from two thousand minute facets.

Can it be? Am I dreaming? Hines Emerald takes its name from its extraordinary green eyes. I blink and turn my head to see it better. Alerted to me now, suddenly wary, the big dragonfly turns its flashing shape at right angles to me, ready to flit away. I see on its thick green thorax what I hoped to see, three yellow diagonal stripes, the front two longer than the third. A charge of adrenalin pumps into my blood, my heart races! It is! It is!

In that instant it is gone. Maybe my pounding heart scared it away. I am left panting, dripping sweat, exhilarated, I feel like leaping and yelling "Yes!" Now my eyes search the bright patch before me again, desperate to see it once more, but there is nothing but the transparent air. Moments pass. I begin questioning whether it was really true, I am doubting, uncertain, wondering if I had just imagined it, wished it into reality. I sit there for long minutes, pondering, hoping that that flash of skittering beauty will return and confirm my memory. It is not to be. The spirits of this magic place allotted me only this one glimpse, no more.

Now something happens that drives the vision of Hines Emerald fleeing in terror from my mind, something just as unforgettable as that rare dragonfly, but still more remarkable.

At first there is a loud crack behind me; it hits me like a plunging knife, then there is a rushing sound that gathers to a roar, then a shriek that rises to a scream that is ear-shattering; it ends with a thunderous thump like a cannon shot that shakes the earth and literally topples me from my sandstone boulder.

For a moment I am frightened beyond reason. Then, as the silence returns and as I look about me in primitive fear, it dawns on me what has happened. That giant beech, which had stood in that woodland glade for three hundred years, had chosen that exact moment to fall crashing to the earth. Why then at that very moment? Why had it gone down when there was no wind at all to push it over? Had a gnat alighted on the side of the tree, its weight just enough to break its last strength and start it falling?

I will never know why. All I will ever know is that that giant tree went down, a hundred feet from me, on a perfectly windless day, in a lonely woods. The shriek and the scream I heard was the spirit of the tree wailing as it was torn from its ancient home. That sound, and that wild rush and thunderous shaking of the earth that followed was an extraordinary gift to me, a gift few men are privileged to receive.

I have been lucky, extraordinarily lucky.

ABOUT THE AUTHOR

Larry was born during World War II in the coal mining town of Linton, Indiana, and grew up there. He was a high school rebel with bad grades and no money, so he had to scratch his way into Vincennes University and on to DePauw University, where he got a B.A. in Chemistry. Then it was on to Manhattan, Kansas for a PhD in Entomology, then to the University of Washington and later Konstanz, Germany. In Germany he was Wissenschaftlicher Assistent to Ernst Florey, the brilliant Austrian comparative physiologist. He and his wife Susie were happy there, thanks in part to becoming the parents of Ian in 1973 and Heather in 1974. In 1975 they felt it necessary to flee back to the USA, fearing still more progeny in Germany. He took a job at the Dept. of Pediatrics at the U of Wisconsin, Madison. After two years there he became Assistant Professor at Purdue, where he made his career in insect physiology and international pest management. He taught a little, probably learning more from his students than they did from him. He has worked a lot in Africa, in extraordinary places like Cameroon, Burkina Faso and Niger, trying to help impoverished farmers store their grain to avoid loss to weevils. He also helped develop transgenic cowpeas with insect resistance. For fun he learned to fly, and now pilots a 1928 Travel Air, an open cockpit biplane built the year his dad graduated from High School. He studies people in the hope of understanding them, writes for pleasure, and takes photographs for the same reason. He's the luckiest guy he knows.

"We are each of us angels with only one wing, and we can only fly by embracing one another."

- Lucretius

www.ingramcontent.com/pod-product-compliance
Lightning Source LLC
Chambersburg PA
CBHW020813300326
41914CB00075B/1716/J